Math Mammoth
Grade 3 Tests and
Cumulative Revisions

for the complete curriculum
(International Series)

Includes consumable student copies of:

- Chapter Tests
- End-of-year Test
- Cumulative Revisions

By Maria Miller

Contents

Grade 3, Chapter 1

End-of-Chapter Test

Instructions to the student:

Answer each question in the space provided. A calculator is not allowed.

Instructions to the teacher:

My suggestion for grading the chapter 1 test is below. The total is 21 points. Divide the student's score by 21 to get a decimal number, and change that decimal to percent to get the student's percentage score.

Question	Max. points	Student score
1	6 points	
2	3 points	
3	3 points	

Question	Max. points	Student score
4	3 points	
5	6 points	
Total	21 points	

Chapter 1 Test

1. Solve using mental maths.

a. $210 + 60 =$ _____	**b.** $55 + 38 =$ _____	**c.** $82 - 35 =$ _____
$198 + 5 =$ _____	$99 + 30 =$ _____	$880 - 9 =$ _____

2. Solve what number goes in place of the triangle.

a. $52 - \triangle = 47$	**b.** $\triangle - 20 = 267$	**c.** $693 + \triangle = 701$
$\triangle =$ _____	$\triangle =$ _____	$\triangle =$ _____

3. Solve.

a. $762 - 756 =$ _____	**b.** $813 - 809 =$ _____	**c.** $355 - 345 =$ _____

4. Write an addition equation and a subtraction equation to match the bar model. Then solve for x.

$$\longmapsto \text{total } \underline{290} \longmapsto$$

130	x

_____ $+$ _____ $=$ _____

_____ $-$ _____ $=$ _____

$x =$ _____

5. Solve. Write a single equation for each problem. Use a **letter for the unknown**.

a. The rent used to be $238, but this month it was raised $9, and next month there will be another $10 increase. How much will the new rent be?

_____

Solution: _____ $=$ _____   The new rent will be $_____ .

b. Ken earned $90 each week for two weeks. Then he paid his parents $60 to help with the cost of food. How much does he have left for himself?

_____

Solution: _____ $=$ _____   He has $_____ left.

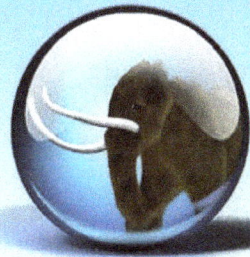

Grade 3, Chapter 2

End-of-Chapter Test

Instructions to the student:

Answer each question in the space provided.

Instructions to the teacher:

My suggestion for grading the chapter 2 test is below. The total is 24 points. Divide the student's score by the total of 24 to get a decimal number, and change that decimal to percent to get the student's percentage score.

Question	Max. points	Student score
1	4 points	
2	4 points	
3a	1 point	
3b	2 points	
3c	2 points	

Question	Max. points	Student score
4	2 points	
5	2 points	
6	4 points	
7	3 points	
Total	24 points	

Chapter 2 Test

1. Round the numbers to the nearest ten.

a. 708 ≈ _____	b. 595 ≈ _____	c. 824 ≈ _____	d. 457 ≈ _____

2. Subtract. *Check* your work.

	Check:		Check:
a. $\begin{array}{r} 4\ 0\ 4 \\ -1\ 5\ 7 \\ \hline \end{array}$		b. $\begin{array}{r} 7\ 2\ 3 \\ -3\ 9\ 7 \\ \hline \end{array}$	

3. **a.** First, estimate the answer to the addition problem.

   57 + 492 + 83 + 146 ≈ _____

   b. Next, calculate to find the exact answer.

   c. Is your answer reasonable?
   How do you know?

4. Jamie has $250. He bought a camera for $127 and batteries for $18.
   Approximately how much money does he have left?

5. One year has 365 days. Of those, 176 are school days.
   How many days in a year are not school days?

6. Calculate.

a. $70 - 40 - 8 + 5 =$ _____

c. $(300 - 30) + (60 - 20) =$ _____

b. $70 - (40 - 8) + 5 =$ _____

d. $300 - 30 + (60 - 20) =$ _____

7. Solve.

$609 - (169 + 145) =$ _____

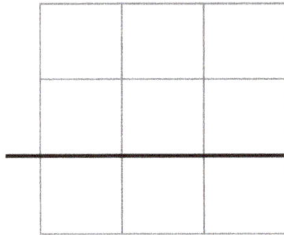
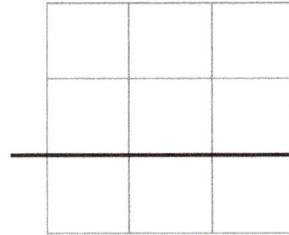

Grade 3, Chapter 3

End-of-Chapter Test

Instructions to the student:

Answer each question in the space provided.

Instructions to the teacher:

My suggestion for grading the chapter 3 test is below. The total is 28 points. Divide the student's score by the total of 28 to get a decimal number, and change that decimal to percent to get the student's percentage score.

Question	Max. points	Student score
1	12 points	
2	4 points	

Question	Max. points	Student score
3	8 points	
4	4 points	
Total	28 points	

Chapter 3 Test

1. Multiply.

a. $2 \times 3 =$ _____	**b.** $2 \times 5 =$ _____	**c.** $2 \times 20 =$ _____	**d.** $1 \times 9 =$ _____
$1 \times 5 =$ _____	$3 \times 10 =$ _____	$3 \times 40 =$ _____	$11 \times 0 =$ _____
$0 \times 7 =$ _____	$2 \times 6 =$ _____	$2 \times 200 =$ _____	$11 \times 1 =$ _____

2. Draw a picture to illustrate the problems.

a. 3×5	**b.** $2 \times 5 + 3 \times 4$

3. Write an equation for each problem and solve.

a. Each basket holds 12 apples. How many apples are in three baskets? Equation: _____ The three baskets have _____ apples.	**b.** Erin had 20 sticks. She made groups of 4 sticks. How many groups did she get? Equation: _____ She got _____ groups of 4 sticks.
c. Chloe bought four pens for $2 each and two games for $8 each. What was the total cost? Equation: _____ The total cost was _____.	**d.** A wall decoration had flowers placed in an array with six rows. In total, it had 30 flowers. How many columns of flowers did it have? Equation: _____ It had _____ columns of flowers.

4. Calculate.

a. $5 + 3 \times 5$	**b.** $20 + 2 \times 3 - 4$
c. $0 \times (10 + 2) \times 3$	**d.** $(8 - 3) \times 1 + 6$

Grade 3, Chapter 4

End-of-Chapter Test

Instructions to the student:

Answer each question in the space provided.

Instructions to the teacher:

Question 1 is to be timed but it's best if you don't let the student know how long a time they have, to avoid anxiety. Allow 3 minutes of time. The student does not have to finish all the questions; there are more questions than what is needed for a perfect score. The point count depends on the number of correct answers as follows: Divide the number of correct answers by 2, and that is the point count, up to 16 points. In other words, a student that gets 32 or more questions correct will get 16 points (no more).

My suggestion for grading the chapter 4 test is below. The total is 38 points. Divide the student's score by the total of 38 to get a decimal number, and change that decimal to percent to get the student's percentage score.

Question	Max. points	Student score
1	16 points	
2	1 point	
3	2 points	
4	2 points	

Question	Max. points	Student score
5	3 points	
6	8 points	
7	6 points	
Total	38 points	

Chapter 4 Test

1. Answer as many questions as you can in the time that your teacher allows. You do not have to finish these.

a.	b.	c.	d.
$2 \times 8 =$ _____	$5 \times 5 =$ _____	$1 \times 8 =$ _____	$6 \times 7 =$ _____
$7 \times 7 =$ _____	$8 \times 4 =$ _____	$6 \times 6 =$ _____	$8 \times 0 =$ _____
$4 \times 6 =$ _____	$5 \times 7 =$ _____	$5 \times 4 =$ _____	$4 \times 4 =$ _____
$5 \times 9 =$ _____	$4 \times 12 =$ _____	$11 \times 4 =$ _____	$9 \times 5 =$ _____

e.	f.	g.	h.
$8 \times 3 =$ _____	$4 \times 8 =$ _____	$8 \times 9 =$ _____	$7 \times 7 =$ _____
$4 \times 6 =$ _____	$3 \times 3 =$ _____	$12 \times 1 =$ _____	$11 \times 8 =$ _____
$5 \times 12 =$ _____	$12 \times 8 =$ _____	$6 \times 11 =$ _____	$6 \times 9 =$ _____
$6 \times 6 =$ _____	$6 \times 5 =$ _____	$9 \times 7 =$ _____	$3 \times 7 =$ _____

i.	j.	k.	l.
$5 \times 2 =$ _____	$3 \times 12 =$ _____	$6 \times 5 =$ _____	$8 \times 7 =$ _____
$8 \times 8 =$ _____	$5 \times 3 =$ _____	$5 \times 8 =$ _____	$12 \times 6 =$ _____
$3 \times 6 =$ _____	$4 \times 5 =$ _____	$7 \times 4 =$ _____	$2 \times 9 =$ _____
$9 \times 4 =$ _____	$6 \times 8 =$ _____	$4 \times 12 =$ _____	$7 \times 6 =$ _____

2. Which multiplication fact is both in the table of 9 and in the table of 7?

3. Find 4×51 using the principle of "doubling twice".

4. Find 17×7 by breaking the multiplication into two parts.

5. Calculate.

a. $3 \times 3 \times 5 =$ _____   **b.** $7 \times 2 \times 5 =$ _____   **c.** $2 \times 9 \times 2 =$ _____

6. Solve. Write a calculation for each problem.

a. A pet store has 10 kittens for sale. Five of them cost $9 each and the rest cost $5 each. How much would all 10 kittens cost?

_____

They would cost _____

b. If one table can seat six people, how many tables do you need for 54 people who are coming to the restaurant?

_____

You need _____ tables.

c. Ann saw seven dogs, four cats, and twelve geese at the park. How many feet do the animals have in total?

_____

They have _____ feet in total.

d. A T-shirt costs $6. How many shirts can you buy with $48?

_____

You can buy _____ T-shirts.

7. Find the missing numbers.

a.	b.	c.	d.
_____ $\times 6 = 24$	$7 \times$ _____ $= 77$	$24 =$ _____ $\times 4$	_____ $\times 8 = 64$
_____ $\times 6 = 54$	$7 \times$ _____ $= 42$	$36 =$ _____ $\times 4$	_____ $\times 8 = 16$
_____ $\times 6 = 36$	$7 \times$ _____ $= 14$	$16 = 4 \times$ _____	$8 \times$ _____ $= 32$

Grade 3, Chapter 5

End-of-Chapter Test

Instructions to the student:

Answer each question in the space provided.

Instructions to the teacher:

My suggestion for grading the chapter 5 test is below. The total is 22 points. Divide the student's score by the total of 22 to get a decimal number, and change that decimal to percent to get the student's percentage score.

Question	Max. points	Student score
1	6 points	
2	2 points	
3	4 points	
4	4 points	

Question	Max. points	Student score
5	2 points	
6	2 points	
7	2 points	
Total	22 points	

Chapter 5 Test

1. Write the time the clock shows, and the time 10 minutes later.

a.   b.   c.

The time
now →

_____ : _____   _____ : _____   _____ : _____

10 min.
later →

_____ : _____   _____ : _____   _____ : _____

2. How much time
   passes from the
   time on the clock
   till the next <u>full hour</u>?

a.   b.

_____ minutes   _____ minutes

3. How much time passes? You can draw a number line for each question to help you.

   a. from 6:46 to 7:21

   b. from 2:52 to 3:14

4. How much time passes?

a. from 4:13 to 6:13 _____ hours	**c.** from 3:10 to 3:53 _____ minutes
b. from 7:30 to 7:50 _____ minutes	**d.** from 11:26 to 12:00 _____ minutes

5. Denny left for orchestra practice at 6:30 PM
   and arrived back home at 9:30 PM.
   How long was he gone?

6. A family left for a vacation on 20 September,
   and returned two weeks later.
   On what date did they return?

September						
Sun	**Mon**	**Tue**	**Wed**	**Thu**	**Fri**	**Sat**
				1	2	3
4	5	6	7	8	9	10
11	12	13	14	15	16	17
18	19	20	21	22	23	24
25	26	27	28	29	30	

7. A soccer game started at 1:30 PM,
   and ended 50 minutes later.
   What time was it then?

Grade 3, Chapter 6

End-of-Chapter Test

Instructions to the student:

Answer each question in the space provided.

Instructions to the teacher:

My suggestion for grading the chapter 6 test is below. The total is 15 points. Divide the student's score by the total of 15 to get a decimal number, and change that decimal to percent to get the student's percentage score.

Question	Max. points	Student score
1	2 points	
2	3 points	
3	4 points	

Question	Max. points	Student score
4	2 points	
5	4 points	
Total	15 points	

Chapter 6 Test

1. How much money? Write the amount.

a. $_____

b. $_____

2. Write as dollar amounts.

two 5-cent coins, three 10-cent coins, and four 20-cent coins	two 50-cent coins, six 10-cent coins, and a 2-dollar coin	four 20-cent coins, four 10-cent coins, and three 1-dollar coins
a. $_____	b. $_____	c. $_____

3. Solve in your head.

   a. You bought stamps for $4.20, a pen for $1.70, and a notebook for $3.30. What was the total cost?

   b. What is your change from $10?

4. Marsha has saved $25, and she wants to buy a game for $41.85. How much does she still need to save?

5. a. Mike bought a sandwich for $4.45, soup for $5.25, juice for $3.65, and water for $2.15.
   Find the total cost.

   b. Find Mike's change from $20.

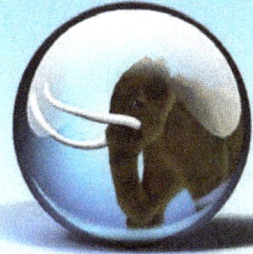

Grade 3, Chapter 7

End-of-Chapter Test

Instructions to the student:

Answer each question in the space provided.

Instructions to the teacher:

My suggestion for grading the chapter 7 test is below. The total is 22 points. Divide the student's score by the total of 22 to get a decimal number, and change that decimal to percent to get the student's percentage score.

Question	Max. points	Student score
1	4 points	
2	4 points	
3	4 points	

Question	Max. points	Student score
4	3 points	
5a	4 points	
5b	3 points	
Total	22 points	

Chapter 7 Test

1. Write each numeral in its proper place to form the number.

a. $2000 + 600 + 80 + 9 =$ _____	**b.** $70 + 4000 =$ _____
c. $600 + 9 + 5000 =$ _____	**d.** $3000 + 2 + 900 =$ _____

2. Compare, and write $<$, $>$, or $=$.

a. $600 + 40$ ☐ $400 + 60 + 1$ **b.** $200 + 7000$ ☐ $5000 + 800$

c. $700 + 5000$ ☐ $50 + 7000$ **d.** $900 + 8$ ☐ $8000 + 9$

3. Find the missing numbers using mental maths.

4. Add $4249 + 512 + 3247 + 77$.

a. $6300 +$ _____ $= 7000$ $9700 - 1500 =$ _____
b. $5400 + 2700 =$ _____ $9000 - 900 =$ _____

5. Solve.

a. An animal park buys animal feed for $1589 and tools for $325. They pay with $2000. What is their change?
b. A new computer costs $2566 and a used one $650. What is the difference in their prices?

Grade 3, Chapter 8

End-of-Chapter Test

Instructions to the student:

Answer each question in the space provided.

Instructions to the teacher:

My suggestion for grading the chapter 8 test is below. The total is 32 points. Divide the student's score by the total of 32 to get a decimal number, and change that decimal to percent to get the student's percentage score.

Question	Max. points	Student score
1	2 points	
2	3 points	
3	12 points	
4	4 points	

Question	Max. points	Student score
5a	2 points	
5b	2 points	
6	5 points	
7	2 points	
Total	32 points	

Chapter 8 Test

<table>
<tr>
<td>

1. Draw a picture to illustrate the division $20 \div 4 = 5$.

</td>
<td>

2. Write a fact family.

_____ × _____ = _____

_____ × _____ = _____

_____ ÷ 6 = 8

_____ ÷ _____ = _____

</td>
</tr>
</table>

3. Divide.

a.	b.	c.	d.
$48 \div 6 =$ _____	$99 \div 11 =$ _____	$49 \div 7 =$ _____	$0 \div 3 =$ _____
$12 \div 3 =$ _____	$70 \div 7 =$ _____	$12 \div 3 =$ _____	$72 \div 6 =$ _____
$96 \div 12 =$ _____	$35 \div 5 =$ _____	$64 \div 8 =$ _____	$18 \div 18 =$ _____

4. Find the missing numbers.

a. _____ $\div 2 = 7$	b. $63 \div$ _____ $= 9$	c. _____ $\div 11 = 10$	d. $7 =$ _____ $\div 8$

5. Write an equation for each problem and solve.

a. Fifty-four children in first grade are arranged into groups of 6 for a trip. How many groups will they make?

Equation: _____

They will make _____ groups.

b. Sammy had 17 crayons and Johnny had 7. They put the crayons together and shared them equally. How many did each boy get?

Equation: _____

Each boy got _____ crayons.

6. Draw a pictograph to show how many baby chicks hatched each week in the hatchery. First, decide how many chicks one chick-picture represents. Write that number in the legend. Optionally, you can also use half of a chick to represent half of that amount.

Chicks Hatched		
Week 8	200	
Week 9	150	
Week 10	225	
Week 11	175	

= _____ chicks

= _____ chicks

7. Refer to question #6 above.

   a. How many more chicks hatched in week 11 than in week 9?

   b. How many chicks were hatched in total during these four weeks?

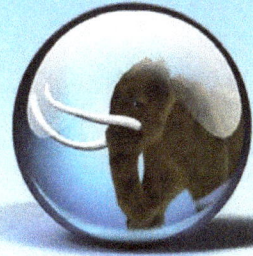

Grade 3, Chapter 9

End-of-Chapter Test

Instructions to the student:

Answer each question in the space provided.

Instructions to the teacher:

My suggestion for grading the chapter 9 test is below. The total is 20 points. Divide the student's score by the total of 20 to get a decimal number, and change that decimal to percent to get the student's percentage score.

Question	Max. points	Student score
1	2 points	
2	2 points	
3	10 points	

Question	Max. points	Student score
4	2 points	
5	2 points	
6	2 points	
Total	20 points	

Chapter 9 Test

1. Draw lines of these lengths:

   a. 95 mm

   b. 6 cm 5 mm

2. Measure the sides of this triangle in millimetres.

3. Fill in each blank with a suitable unit.

a. Mary's book weighed 350 _____.	**f.** Mum bought 3 _____ of bananas.
b. A box of juice had 2 _____ of juice.	**g.** Erika weighs 55 _____.
c. The aeroplane was flying 10 000 _____ above the ground.	**h.** A cell phone weighs 120 _____.
	i. A housefly measured 7 _____ long.
d. The large tank holds 200 _____ of water.	**j.** The shampoo bottle can hold
e. Andy and Matt bicycled 10 _____ to the beach.	450 _____ of shampoo.

4. A T-shirt has a mass of 200 g. How many of them, placed on a scale, would make the scale show 1 kilogram (1 kg = 1000 grams)?

5. A bucket contains 16 litres of water. Sheila pours equal amounts of water out of it into four containers. How much water is in each container?

6. How much liquid is in each pitcher?

a.

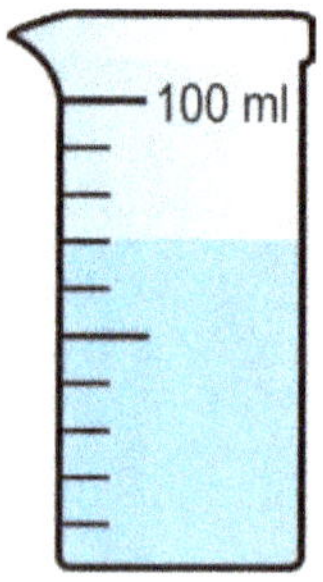

100 ml

_____ ml

b.

800 ml

_____ ml

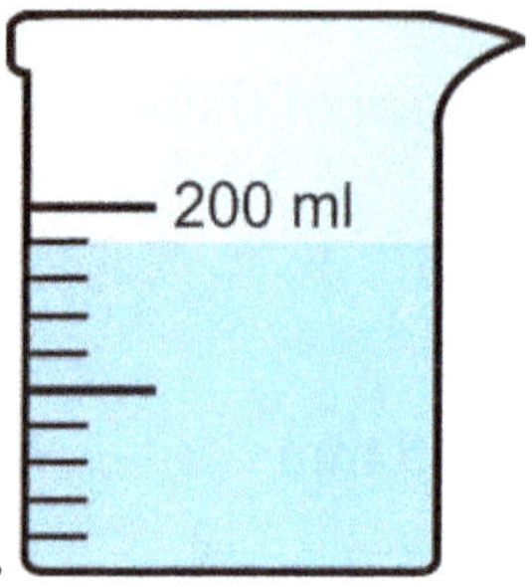

c.

200 ml

_____ ml

Grade 3, Chapter 10

End-of-Chapter Test

Instructions to the student:

Answer each question in the space provided.

Instructions to the teacher:

My suggestion for grading the chapter 10 test is below. The total is 25 points. Divide the student's score by the total of 25 to get a decimal number, and change that decimal to percent to get the student's percentage score.

Question	Max. points	Student score
1	7 points	
2	2 points	
3	2 points	
4	4 points	

Question	Max. points	Student score
5	2 points	
6	3 points	
7	2 points	
8	3 points	
Total	25 points	

Chapter 10 Test

1. Match each shape to its description.

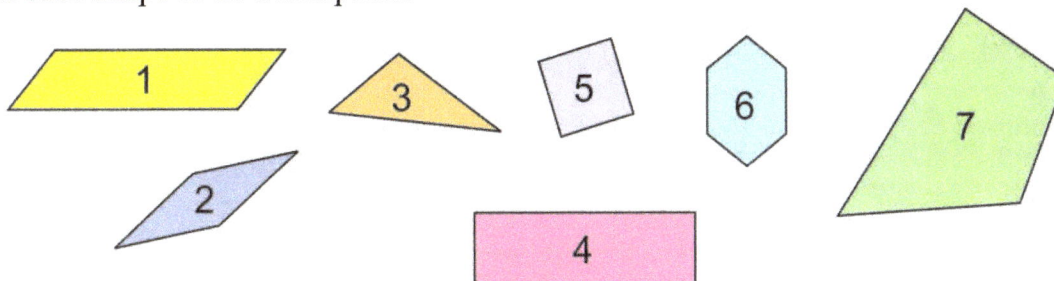

a. It is a rhombus.

b. It has no equal sides.

c. It has four equal sides and four right angles.

d. It has two pairs of equal sides and no right angles.

e. It has two pairs of equal sides, and four right angles.

f. It is a hexagon.

g. It is a quadrilateral, and has no equal sides.

2. Sketch here...

a. a rhombus	**b.** a quadrilateral that is not a rectangle nor a rhombus

3. Find the area and perimeter of this figure.

Area = _____

Perimeter = _____

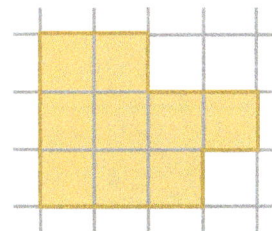

4. Find the area and perimeter of these rectangles.

a. 2 m, 4 m

Perimeter = _____

Area = _____

b. 7 cm, 7 cm

Perimeter = _____

Area = _____

5. Write an expression with two multiplications for the total area of this shape, in square units.

____ × ____ + ____ × ____ = _____

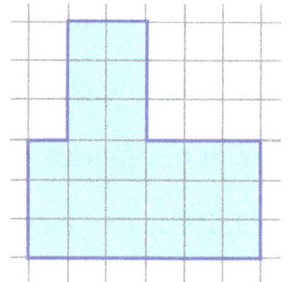

6. Samantha's lawn is in the L-shape shown on the right. Calculate its area.

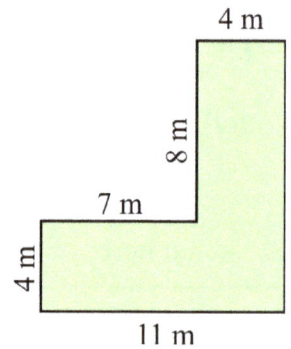

4 m

8 m

7 m

4 m

11 m

7. Jorge is planning to build a pen for his sheep. One possible pen would be an 18 m by 24 m rectangle, and the other possible pen would be a 12 m by 36 m rectangle.
Which pen has a larger perimeter? How much larger?

8. Draw a two-part rectangle that matches the given expression, and fill in the missing parts.

____ × (____ + ____) = 4 × 2 + 4 × 5

Grade 3, Chapter 10

End-of-Chapter Test, Version 2

Instructions to the student:

Answer each question in the space provided.

Instructions to the teacher:

My suggestion for grading the chapter 10 test is below. The total is 25 points. Divide the student's score by the total of 25 to get a decimal number, and change that decimal to percent to get the student's percentage score.

Question	Max. points	Student score
1	6 points	
2	1 point	
3	2 points	
4	2 points	
5	4 points	

Question	Max. points	Student score
6	2 points	
7	3 points	
8	2 points	
9	3 points	
Total	25 points	

Chapter 10 Test (Version 2)

1. Match each shape to its description.

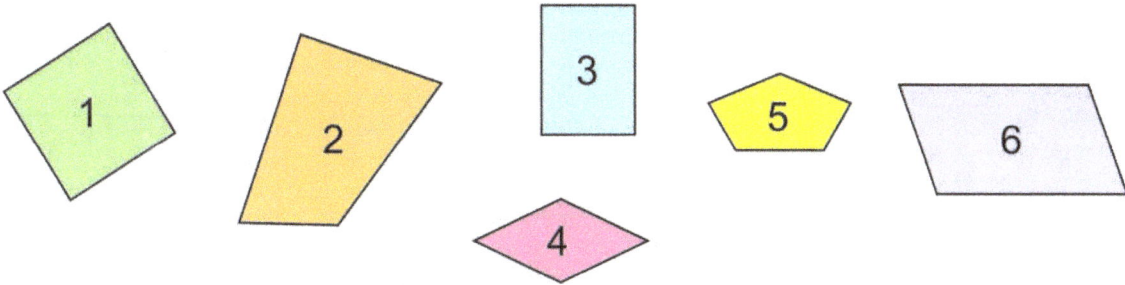

   a. It has four equal sides and no right angles.

   b. It has four equal sides and four right angles.

   c. It is a pentagon.

   d. It has two pairs of equal sides and no right angles.

   e. It has four right angles and not all its sides are equal.

   f. It is a quadrilateral, and has no equal sides.

2. Sketch here a quadrilateral that has two equal sides, but no right angles.

3. Find the area and perimeter of this figure.

   Area = _____

   Perimeter = _____

4. The perimeter of this rectangle is 108 cm. Its one
   side is 21 cm. How long is the other side?

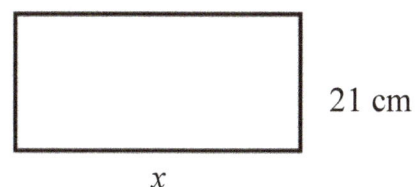

21 cm

x

5. Find the area and perimeter of these rectangles.

a.

3 m

5 m

Perimeter = _____

Area = _____

b.

9 cm

9 cm

Perimeter = _____

Area = _____

6. Write an expression with two multiplications for the total area of this shape, in square units.

___ × ___ + ___ × ___ = _____

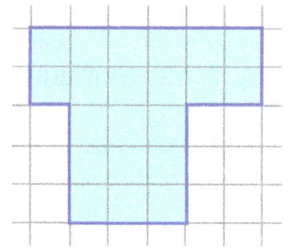

7. The floor of a building is in the L-shape shown on the right. Calculate its area.

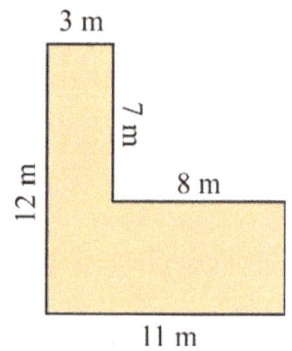

3 m

7 m

12 m

8 m

11 m

8. Lindsay has a rectangular 5 m by 7 m garden in her back yard. Her neighbour also has a garden, 3 m by 8 m.

Which garden has a larger perimeter?

How much larger?

9. Write a number sentence for the total area, thinking of one rectangle or two.

___ × (___ + ___) = ___ × ___ + ___ × ___

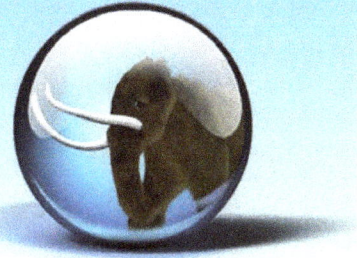

Grade 3, Chapter 11

End-of-Chapter Test

Instructions to the student:

Answer each question in the space provided.

Instructions to the teacher:

My suggestion for grading the chapter 11 test is below. The total is 27 points. Divide the student's score by the total of 27 to get a decimal number, and change that decimal to percent to get the student's percentage score.

Question	Max. points	Student score
1	3 points	
2	2 points	
3	5 points	
4	4 points	
5	4 points	

Question	Max. points	Student score
6	2 points	
7	3 points	
8	2 points	
9	2 points	
Total	27 points	

Chapter 11 Test

1. Divide each shape into parts and shade parts to illustrate each fraction.

a. $\dfrac{3}{4}$ **b.** $\dfrac{3}{2}$ **c.** $\dfrac{9}{8}$

2. Mark the given fraction on the number line.

a. $\dfrac{1}{4}$ 0 ————————— 1

b. $\dfrac{2}{6}$ 0 ————————— 1

3. Mark the fractions on the number line: $\dfrac{9}{5}$, $\dfrac{13}{5}$, $\dfrac{6}{5}$, $\dfrac{2}{5}$, $\dfrac{15}{5}$.

```
0               1               2               3
├───────────────┼───────────────┼───────────────┤
```

4. Write the whole numbers as fractions.

a. $1 = \underline{\quad}$ **b.** $2 = \underline{\quad}$ **c.** $4 = \dfrac{\quad}{8}$ **d.** $5 = \dfrac{\quad}{3}$

5. Compare the fractions, and write >, <, or = .

a. $\dfrac{8}{10}$ ☐ $\dfrac{9}{10}$ **b.** 1 ☐ $\dfrac{3}{7}$ **c.** $\dfrac{2}{3}$ ☐ $\dfrac{2}{5}$ **d.** $\dfrac{6}{3}$ ☐ $\dfrac{6}{4}$

6. Write the fractions $\dfrac{1}{3}$, $\dfrac{1}{6}$, and $\dfrac{2}{2}$ in order from the smallest to the greatest.

$$\underline{\quad} < \underline{\quad} < \underline{\quad}$$

7. Write and shade the equivalent fractions.

a. $\dfrac{2}{5}$ =

b. $\dfrac{6}{8}$ =

c. $\dfrac{2}{3}$ = ——

8. A loaf of bread is cut into 12 equal pieces.
   Another, similar loaf, is cut into 8 pieces.

   Mike ate 3 pieces of the first loaf. Janet ate 2 pieces
   of the second loaf. Who ate more bread? Explain.

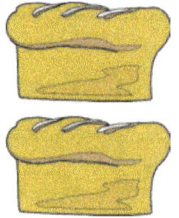

9. This picture is trying to show that $\dfrac{3}{9} = \dfrac{3}{8}$.

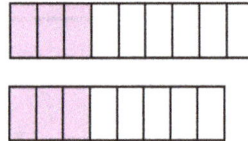

   Explain why it is wrong.

   $$\dfrac{3}{9} = \dfrac{3}{8}$$

End-of-Year Test - Grade 3

This test is quite long, so I do not recommend having your child/student do it in one sitting. Break it into parts and administer them either on consecutive days, or perhaps on morning/evening/morning. This is to be used as a diagnostic test. You may even skip those areas that you already know for sure your student has mastered.

The test does not cover every single concept that is covered in the *Math Mammoth Grade 3 International Version,* but all the major concepts and ideas are tested here. This test is evaluating the child's ability in the following content areas:

- multiplication tables and basic division facts
- mental maths
- regrouping in addition and subtraction; checking subtraction with addition
- rounding to the nearest ten
- basic word problems
- writing an equation for a word problem
- estimating
- multiplication and related concepts
- clock to the minute and elapsed time calculations
- reading a bar graph; making a pictograph
- basic money calculations (finding totals and change)
- place value with four-digit numbers
- adding and subtracting four-digit numbers
- division and related concepts (division by 1 or 0, word problems)
- measuring lines in inches and centimetres
- basic usage of measuring units
- attributes of shapes
- area and perimeter of rectangles
- the concept of a fraction, fractions on a number line
- equivalent fractions
- comparing fractions

Note 1: problems #2 and #3 are done <u>orally and timed</u>. Let the student see the problems. Read each problem aloud, and wait a maximum of 5-6 seconds for an answer. Mark the problem as right or wrong according to the student's (oral) answer. Mark it wrong if there is no answer. Then you can move on to the next problem.

You do not have to mention to the student that the problems are timed or that they will have 5-6 seconds per answer, because the idea here is not to create extra pressure by the fact it is timed, but simply to check if the student has the facts memorised (quick recall). You can say for example (vary as needed):

"I will ask you some multiplication and division questions. Try to answer me as quickly as possible. In each question, I will only wait a little while for you to answer, and if you do not say anything, I will move on to the next problem. So just try your best to answer the questions as quickly as you can."

In order to continue with the Math Mammoth Grade 4 Complete Curriculum, I recommend that the child gain a minimum score of 80% on this test, and that the teacher or parent revise with him any content areas that are found weak. Children scoring between 70 and 80% may also continue with grade 4, depending on the types of errors (careless errors or not remembering something, vs. lack of understanding). The most important content areas to master are the multiplication tables and the word problems, because of the level of logical reasoning needed in them. Use your judgment.

Instructions to the student: Answer each question in the space provided.

Instructions to the teacher: My suggestion for grading is below. The total is 219 points. A score of 175 points is 80%.

Grading on question 1 (the multiplication tables grid): There are 144 empty squares to fill in the table, and the completed table is worth 14 points. Count how many of the answers the student gets right, divide that by 10, and round to the nearest whole point. For example: a student gets 24 right. 24/ 10 = 2.4, which rounded becomes 2 points. Or, a student gets 85 right. 85/10 = 8.5, which rounds to 9 points.

Grading on question 2: Each question is worth 1/2 point.

Question	Max. points	Student score
Multiplication Tables and Basic Division Facts		
1	14 points	
2	8 points	
3	8 points	
	subtotal	/ 30
Addition and Subtraction		
4	6 points	
5	6 points	
6	3 points	
7	2 points	
8	3 points	
	subtotal	/ 20
Regrouping and Rounding		
9	3 points	
10	2 points	
11	4 points	
12	3 points	
13	4 points	
14	3 points	
	subtotal	/ 19

Question	Max. points	Student score
Multiplication and Related Concepts		
15	1 point	
16	1 point	
17	3 points	
18	3 points	
19	3 points	
20a	2 points	
20b	2 points	
20c	2 points	
20d	2 points	
	subtotal	/ 19
Time		
21	6 points	
22	2 points	
23	4 points	
24	4 points	
	subtotal	/ 16
Graphs		
25	4 points	
26	3 points	
	subtotal	/ 7

Question	Max. points	Student score
Money		
27	4 points	
28	3 points	
29	3 points	
	subtotal	/ 10
Four-Digit Numbers		
30	2 points	
31	2 points	
32	5 points	
33	4 points	
34	4 points	
	subtotal	/ 17
Division and Related Concepts		
35	2 points	
36	9 points	
37	6 points	
38	6 points	
	subtotal	/ 23
Measuring		
39	2 point	
40	1 point	
41	1 point	
42	1 point	
43	6 points	
	subtotal	/ 11

Question	Max. points	Student score
Geometry		
44	6 points	
45	3 points	
46	2 points	
47	3 points	
48	2 points	
49	2 points	
50	2 points	
	subtotal	/ 20
Fractions		
51	5 points	
52	5 points	
53	4 points	
54	3 points	
55	2 points	
56	5 points	
57	3 points	
	subtotal	/ 27
TOTAL		/ 219

Math Mammoth Grade 3 International Version
End-of-Year Test

Multiplication Tables and Basic Division Facts

1. Your first problem will be to fill in the complete multiplication table.
   You have 12 minutes to fill it in completely.

×	1	2	3	4	5	6	7	8	9	10	11	12
1												
2												
3												
4												
5												
6												
7												
8												
9												
10												
11												
12												

In problems 2 and 3, your teacher will read you multiplication and division questions. Try to answer them as quickly as possible. In each question, he/she will only wait a little while for you to answer, and if you do not say anything, your teacher will move on to the next problem. So just try your best to answer the questions as quickly as you can.

2. Multiply.

a.	b.	c.	d.
$2 \times 7 =$ _____	$7 \times 4 =$ _____	$3 \times 3 =$ _____	$7 \times 8 =$ _____
$8 \times 3 =$ _____	$5 \times 8 =$ _____	$4 \times 4 =$ _____	$6 \times 5 =$ _____
$5 \times 5 =$ _____	$3 \times 9 =$ _____	$7 \times 7 =$ _____	$8 \times 6 =$ _____
$9 \times 4 =$ _____	$5 \times 7 =$ _____	$4 \times 8 =$ _____	$6 \times 9 =$ _____

3. Divide.

a.	b.	c.	d.
$21 \div 3 =$ _____	$32 \div 4 =$ _____	$45 \div 5 =$ _____	$50 \div 5 =$ _____
$35 \div 7 =$ _____	$40 \div 8 =$ _____	$28 \div 4 =$ _____	$72 \div 9 =$ _____
$48 \div 6 =$ _____	$66 \div 6 =$ _____	$36 \div 9 =$ _____	$18 \div 6 =$ _____
$49 \div 7 =$ _____	$56 \div 8 =$ _____	$63 \div 7 =$ _____	$27 \div 9 =$ _____

Addition and Subtraction

4. Add in your head and write the answers.

a. 240 + 70 = _____	**b.** 540 + 80 = _____	**c.** 59 + 89 = _____
99 + 50 = _____	335 + 9 = _____	46 + 34 = _____

5. Subtract in your head and write the answers.

a. 100 – 67 = _____	**b.** 651 – 8 = _____	**c.** 52 – 37 = _____
73 – 68 = _____	54 – 9 = _____	400 – 22 = _____

6. Write one addition and two subtraction
   equations to match the bar model.

   Write the numbers in the bar model, also.

$$\vdash \text{———— total } \boxed{} \text{ ————} \dashv$$

_____ + _____ = _____

900 – _____ = 440

_____ – _____ = _____

7. A family is driving 300 km from their hometown to Grandma's place.
   Ten kilometres before the half-way point they stop to have lunch.
   How many kilometres do they still have to go?

8. Write an equation (or several) for the problem. Use a letter for the unknown.

A store owner had a fridge with a price of $400. Then he doubled the price.
A customer came, and he told the customer, "I will take some money off the price."
So, the customer paid $600. How much did the dealer take off the price?

Equation: _____

Solution: _____

Regrouping and Rounding

9. Round these numbers to the nearest ten.

| a. 93 ≈ _____ | b. 607 ≈ _____ | c. 455 ≈ _____ |

10. Each week, Eric puts $38 in his savings. He wants to purchase a blender for $189. Round the numbers, and then estimate in how many weeks he could buy the blender.

11. Solve what number goes in place of the triangle.

a. $414 + \triangle = 708$

$\triangle$ is _____

b. $\triangle - 339 = 485$

$\triangle$ is _____

12. Solve. Check that your final answer is reasonable.

Jason wants to buy a $545 camera and a $52 camera bag. Right now, he has saved $310. Your task is to find out how much more money he needs.

a. Write an equation for this problem that uses a letter for the unknown.

_____

b. Estimate the final answer: Jason needs about $_____ more.

c. Now calculate the exact answer. Jason needs $_____ more.

13. Subtract. Then check your answer.

a.   Check:

$$\begin{array}{r} 9\ 6\ 2 \\ -\ 3\ 8\ 3 \\ \hline \end{array}$$

b.   Check:

$$\begin{array}{r} 7\ 0\ 3 \\ -\ 5\ 4\ 6 \\ \hline \end{array}$$

14. The calculation on the right shows how
    Joe added $82 + 539 + 154 + 8$.

    a. *Estimate* the result of $82 + 539 + 154 + 8$.

    b. How could Joe see that his answer is not
    reasonable?

    c. Correct the error Joe made.

$$\begin{array}{r} {\scriptstyle 2}\quad\quad \\ 8\ 2 \\ 5\ 3\ 9 \\ 1\ 5\ 4 \\ +\quad\quad 8 \\ \hline 8\ 5\ 5 \end{array}$$

Multiplication and Related Concepts

15. Draw a picture to illustrate
    the multiplication $3 \times 4 = 12$.

16. Solve: $3 \times 25 =$ _____

17. Solve.

| **a.** $80 \times 3 =$ _____ | **b.** $7 \times 70 =$ _____ | **c.** $6 \times 50 =$ _____ |

61

18. Write a multiplication equation (NOT just the answer) to solve how many legs these animals have in total.

  a. seven horses _____

  b. five ducks _____

  c. eight horses and six ducks _____

19. Solve.

a. $24 + 8 \times 3$	**b.** $2 + (5 + 4) \times 2$	**c.** $66 - 5 \times 5$

20. Solve. Write an equation for each problem.

a. Pat wants to have three rolls for each of her 12 guests. How many rolls does she need?

Equation: _____   She needs _____ rolls.

b. Each table in a restaurant seats four people. How many tables do you need to reserve for a party of 32 people?

Equation: _____   You need _____ tables.

c. A cafeteria menu had spaghetti with meatballs for $11 and bean soup for $8. How much would it cost to buy three plates of spaghetti with meatballs and three bowls of bean soup?

Equation: _____   It would cost $_____.

d. Anna is bagging hair clips she made. She will put four hair clips in each bag. She has 28 hair clips to bag. How many bags will she need?

Equation: _____   She will need _____ bags.

Time

21. Write the time the clock shows. Below, write the time 10 minutes later.

a.	b.	c.

The time
now →   _____ : _____   _____ : _____   _____ : _____

10 min.
later →   _____ : _____   _____ : _____   _____ : _____

22. How many minutes is it
from the time on the clock face
until the given time?

until 1:00   until 5:55

a. _____ minutes   b. _____ minutes

23. How much time passes? You can draw a number line for each question to help you.

    a. from 2:49 AM to 3:12 AM

    b. from 11:33 AM to 12:06 PM

24. a. Jess started to watch an animal video at 4:35 PM
    and she stopped at 4:52 PM. How long did she watch it?

    b. A casserole dish needs to bake for 45 minutes. If it needs to be
    ready at 6:30 PM, when should it go to the oven?

Graphs

25. The graph shows some people's working hours on Uncle Ted's apple farm. **Each block means 5 hours of work.**

    a. How many hours did Chloe and Kathy work in total?

    b. How many more hours did Jason work than Jack?

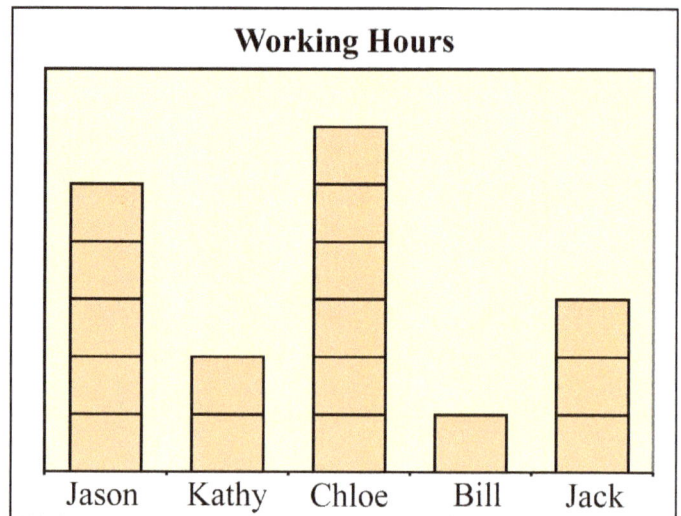

Working Hours

Jason   Kathy   Chloe   Bill   Jack

    c. How many less hours did Bill work than Jack?

    d. How many hours did Jason, Bill, and Jack work in total?

26. The table below lists how many hours of tennis practice each person did in a month. Make a pictograph from the data. Use a tennis ball for the picture. Choose how many hours each tennis ball picture represents.

Tennis Practice (hours)	
Ava	6
Juan	3
Greg	9
Adelaide	12

Tennis Practice	
Ava	
Juan	
Greg	
Adelaide	

= _____ hours

Money

27. Solve using mental maths.

a. You buy a book for $15.10 and stickers for $3.50. You give $20. Your total: $_____ Change: $_____	**b.** You buy two baskets for $14.50 each. You give $30. Your total: $_____ Change: $_____

28. Find the total cost of buying the items listed. Line up the numbers carefully when you add.

$19.90	$12.95	$3.25	$36.59

a. a stuffed elephant and a bag	**b.** two pens and a book	**c.** three stuffed elephants

29. A pencil case costs $7.85. If Mark buys three of them and pays with $50, what will his change be?

Four-Digit Numbers

30. These numbers are written as sums. Write them in the normal way.

a. 5000 + 200 + 5 = _____	**b.** 90 + 2000 + 4 = _____
c. 300 + 7000 = _____	**d.** 2 + 8000 = _____

31. Fill in the missing part.

a. 2000 + 60 + _____ = 2760	**b.** 700 + 20 + _____ + 9 = 2729

32. Compare and write < , > , or = in the box.

a. 6034 ☐ 3064	**b.** 5156 ☐ 5516	**c.** 9079 ☐ 9097
d. 6000 + 3 + 40 ☐ 400 + 60 + 3000	**e.** 900 + 7000 ☐ 90 + 7000 + 2	

33. Add and subtract.

a. 5400 + 300 = _____ 7800 + 800 = _____	**b.** 2900 − 1700 = _____ 8100 − 300 = _____

34. Solve. Check your work by adding.

a.

$$\begin{array}{r} 8\ 1\ 4\ 9 \\ -\ 2\ 8\ 8\ 8 \\ \hline \end{array} \qquad + \underline{\hspace{3cm}}$$

b.

$$\begin{array}{r} 6\ 4\ 3\ 6 \\ -\ 3\ 7\ 4\ 9 \\ \hline \end{array} \qquad + \underline{\hspace{3cm}}$$

Division and Related Concepts

35. Write two multiplications and two divisions for the same picture.

_____ × _____ = _____   _____ ÷ _____ = _____

_____ × _____ = _____   _____ ÷ _____ = _____

36. Fill in the missing numbers.

a. 50 ÷ _____ = 5	**b.** 6 × _____ = 48	**c.** 64 ÷ _____ = 8
d. 5 = 45 ÷ _____	**e.** 20 = 4 × _____	**f.** _____ ÷ 5 = 8
g. _____ × 9 = 54	**h.** _____ ÷ 9 = 12	**i.** 32 ÷ _____ = 4

37. Divide, but CROSS OUT all the problems that are impossible!

a. 17 ÷ 1 = _____ 17 ÷ 0 = _____	**b.** 17 ÷ 17 = _____ 0 ÷ 0 = _____	**c.** 1 ÷ 1 = _____ 0 ÷ 1 = _____

38. Solve. Write an equation for each problem, using a letter for the unknown.

a. Camila, Leo, and Daniel decided to buy a gift that cost $16 and flowers that cost $14 for Mum. The children shared the total cost equally. How much did each child pay?

_____   Each child paid $_____.

b. Each minibus holds ten passengers. There are six full minibuses, and one with one empty seat. How many passengers are there in total?

_____   There are _____ passengers.

c. The Smith family made 24 sandwiches for a picnic. They packed them in containers, six sandwiches in each. How many containers did they use?

_____   They used _____ containers.

67

Measuring

39. Draw lines of these lengths:

    a. 11 cm 2 mm

    b. 75 mm

40. Write in order from the smallest to the biggest unit: *cm   km   m   mm*

41. Anna put three apples on a scale, one at a time.
    The scale read 120 g, 105 g, and 130 g.
    What is the total mass of her apples?

42. Samuel filled three 4-litre pitchers with water.
    How much water is in those pitchers, in total?

43. Fill in the blanks with units of measure. Sometimes several different units are possible.

    a. The family drove 50 _____ to visit a beach.   **b.** The pencil is 14 _____ long.

    c. Two paperclips weigh 5 _____ .   **d.** The teacher weighs 68 _____ .

    e. The large glass holds 300 _____ of liquid.   **f.** A dropper measures 2 _____ .

Geometry

44. Match each description to one shape.

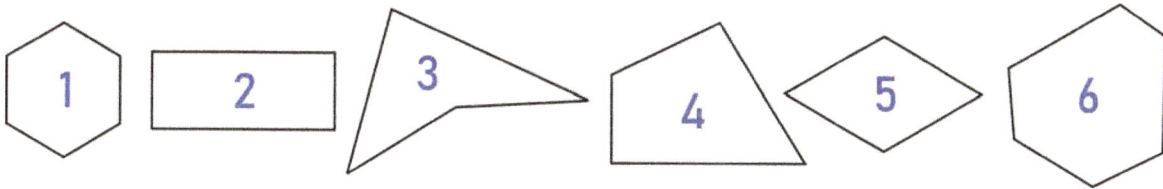

    a. A quadrilateral with four right angles.

    b. A quadrilateral with no right angles.

    c. A regular hexagon.

    d. A quadrilateral with exactly one right angle.

    e. A hexagon that is not regular.

    f. A rhombus.

45. Sketch here:

    a. A triangle that doesn't have any equal sides.

    b. A quadrilateral that has four right angles,
       and not all its sides are equal.

    c. A rhombus that doesn't have any right angles.

46. Find the perimeter and area of this shape.

    Perimeter: _____

    Area : _____

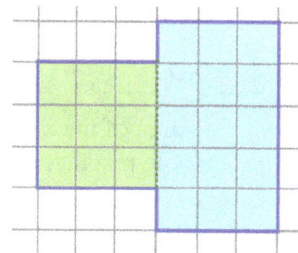

47. The picture shows a two-part lawn that is made
    of two rectangles.

    a. Find the areas of rectangles 1 and 2.

    _____ and _____

    b. Find the perimeter of the whole lawn.

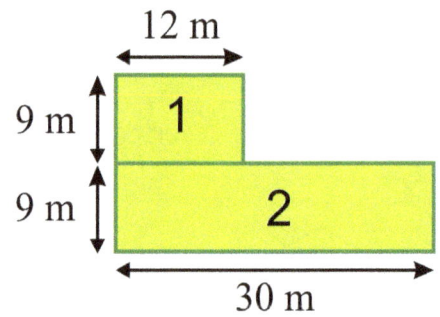

12 m

9 m 1

9 m 2

30 m

48. The perimeter of a rectangle measures 26 cm.
    Find the other side length, if one side measures 4 cm.

49. Draw in the grid below:

    a. a rectangle with an area of 15 square units

    b. a rectangle with a perimeter of 10 units.

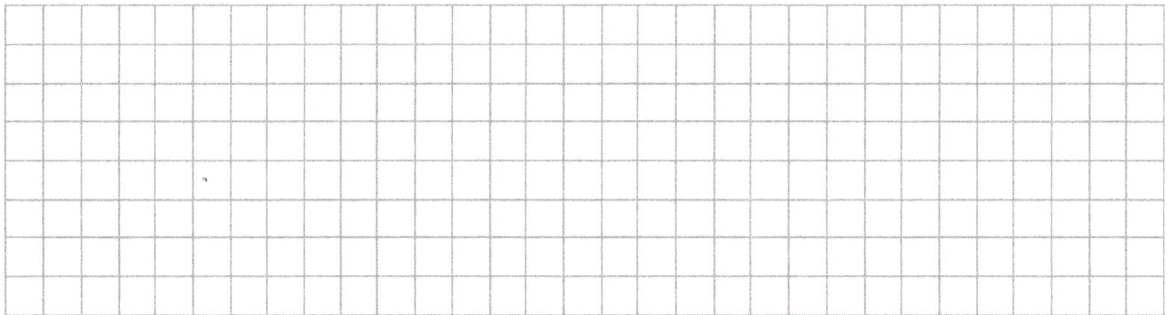

50. Write a number sentence for the total area, thinking of one rectangle or two.

____ × (____ + ____) = ____ × ____ + ____ × ____ = _____

  area of the   area of the   area of the
whole rectangle   first part   second part

Fractions

51. Write the fraction.

a. _____   b. _____   c. _____

d. _____   e. _____

52. Mark these fractions on the number line: $\frac{5}{3}$, $\frac{8}{3}$, $\frac{9}{3}$, $\frac{11}{3}$, $\frac{3}{3}$.

53. Find the fractions that are equal to some whole number.

a. $\frac{6}{4}$	b. $\frac{8}{8}$	c. $\frac{8}{2}$	d. $\frac{2}{8}$	e. $\frac{13}{3}$	f. $\frac{24}{4}$	g. $\frac{27}{3}$	h. $\frac{20}{6}$

54. Write an equivalent fraction, based on the illustration.

a. $\frac{3}{4}$ = _____   b. $\frac{10}{12}$ = _____   c. $\frac{2}{3}$ = _____

55. Show that the fractions $\frac{3}{6}$ and $\frac{1}{2}$ are equivalent, using the number lines.

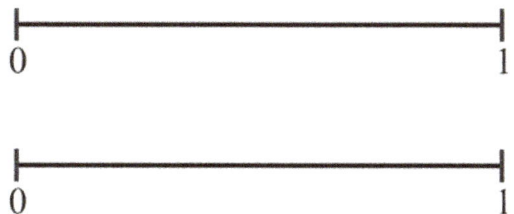

56. Compare the fractions, and write $>$, $<$, or $=$ in the box.

a. $\dfrac{2}{8}$ ☐ $\dfrac{2}{3}$ b. $\dfrac{5}{10}$ ☐ $\dfrac{7}{10}$ c. $\dfrac{6}{3}$ ☐ 2 d. $\dfrac{1}{6}$ ☐ $\dfrac{1}{8}$ e. $\dfrac{3}{6}$ ☐ $\dfrac{1}{2}$

57. Compare the fractions, writing $>$, $<$, or $=$ between them. If you cannot make a valid comparison, then cross the whole problem out.

a.	b.	c.
$\dfrac{3}{10}$ $\dfrac{2}{8}$	$\dfrac{3}{9}$ $\dfrac{2}{6}$	___ ___

Using the cumulative revisions and the worksheet maker

The cumulative revisions contain a mix of problems, practising topics from the curriculum up to the chapter named in the revision. For example, the cumulative revision for chapters 1-6 may include problems matching chapters 1, 2, 3, 4, 5, and 6. It can be used any time after the student has studied the curriculum through chapter 6.

These cumulative revision lessons are optional; use them as needed. The student doesn't have to complete all of them, however I recommend using at least three of these revisions during the school year. The teacher can also use the revisions as diagnostic tests to find out what topics the student has trouble with.

Math Mammoth complete curriculum also includes an easy worksheet maker, which is the perfect tool to make more problems for children who need more practice. The worksheet maker covers most topics in the curriculum, though usually not many word problems. Most people find it to be a very helpful addition to the curriculum.

You can also access the worksheet maker online at

https://www.mathmammoth.com/private/Make_extra_worksheets_grade3.htm

In addition to the cumulative revisions and the worksheet maker, we also offer a free online practice area at https://www.mathmammoth.com/practice/. This section of the website has a growing number of games and practice activities for many maths topics.

Cumulative Revision, Grade 3, Chapters 1-2

1. Add. Look for the easiest order to add.

    a. $31 + 40 + 2 + 9$ **b.** $5 + 50 + 20 + 8$ **c.** $22 + 61 + 4 + 8$

    = _____   = _____   = _____

    d. $15 + 30 + 15 + 6$ **e.** $52 + 8 + 20 + 3$ **f.** $8 + 11 + 9 + 7$

    = _____   = _____   = _____

2. Add and subtract in your head.

a. $560 + 40 =$	**b.** $730 + 80 =$	**c.** $990 - 80 =$
d. $230 - 70 =$	**e.** $610 - 60 =$	**f.** $900 - 660 =$

3. Round the numbers to the nearest ten.

a. $506 \approx$ _____	**b.** $757 \approx$ _____	**c.** $904 \approx$ _____	**d.** $8 \approx$ _____

4. Estimate the costs using rounded numbers.

a. rent, \$256, and groceries, \$387	**b.** 1 adult's ticket, \$58, and 1 child's ticket, \$38
rent about \$_____	adult's ticket about \$_____
groceries about \$_____	child's ticket about \$_____
total about \$_____	total cost about \$_____

5. Calculate in the correct order.

a. $35 - 14 - 7 + 3 =$ _____	**d.** $(250 - 20) + (80 - 30) =$ _____
b. $35 - (14 - 7) + 3 =$ _____	**e.** $250 - (20 + 80 - 30) =$ _____
c. $35 - (14 - 7 + 3) =$ _____	**f.** $250 - 20 + (80 - 30) =$ _____

6. Find patterns in this addition table, and use the patterns to fill it in without having to individually add each pair of numbers.

+	40	43	46	49	52
32					
34					
36					
38					

7. Write one addition and one subtraction sentence to match the bar model.
   Also, write the numbers in the model.

|—— total _____ ——|

a. 250 + _____ = 400

_____ − _____ = _____

|—— total _____ ——|

b. _____ + _____ = _____

500 − _____ = 390

8. Write an equation (or several) for each problem. Use a letter for the unknown.

a. Jasmine has a tea set with 14 cups and a tea set with 13 cups. She wants to invite the girls from her class for a tea party. There are 30 girls in her class. How many more cups does she need?

Equation: _____

Solution: _____ = _____

She needs _____ more cups.

b. Seth had $150. Then he bought clothes for $60. After that, he earned $40. How much money does he have now?

Equation: _____

Solution: _____ = _____

He has $_____ now.

76

Cumulative Revision, Grade 3, Chapters 1 - 3

1. Three parts make up one whole. Write an addition equation to match the bar model. Then solve for the unknown.

total 698
x	196	153

_____ + _____ + _____ = _____

$x =$ _____

2. Subtract. Check your work.

a.
$$
\begin{array}{r}
8\ 8\ 8 \\
-\ 2\ 9\ 9 \\
\hline
\end{array}
\qquad +\ \underline{\hspace{2cm}}
$$

b.
$$
\begin{array}{r}
4\ 5\ 0 \\
-\ 1\ 3\ 4 \\
\hline
\end{array}
\qquad +\ \underline{\hspace{2cm}}
$$

c.
$$
\begin{array}{r}
6\ 0\ 2 \\
-\ 3\ 4\ 4 \\
\hline
\end{array}
\qquad +\ \underline{\hspace{2cm}}
$$

d.
$$
\begin{array}{r}
8\ 0\ 0 \\
-\ 6\ 5\ 7 \\
\hline
\end{array}
\qquad +\ \underline{\hspace{2cm}}
$$

3. Can you buy three bicycles for $96 each and pay with $300?

   If yes, how much money will you have left?

   If not, how much more money would you need?

4. Multiply.

a.	b.	c.	d.
$3 \times 3 =$ _____	$5 \times 2 =$ _____	$3 \times 30 =$ _____	$10 \times 0 =$ _____
$0 \times 8 =$ _____	$3 \times 5 =$ _____	$2 \times 400 =$ _____	$22 \times 1 =$ _____

5. Write < , > , or = .

   a. $350 - 18$ ☐ $350 - 15$ **b.** $180 - 15$ ☐ $190 - 25$ **c.** $264 + 7$ ☐ $267 + 8$

6. Subtract.

a. $644 - 8 =$ _____	**b.** $277 - 9 =$ _____	**c.** $683 - 8 =$ _____
$233 - 7 =$ _____	$191 - 5 =$ _____	$842 - 7 =$ _____

7. The calculation on the right shows how
   Grace added $186 + 377 + 29 + 82$.

   a. *Estimate* the result of $186 + 377 + 29 + 82$.

   b. How could Grace see that her answer is not
      reasonable?

```
      3 1
    1 8 6
    3 7 7
      2 9
  +   8 2
  ─────────
    9 3 5
```

   c. Correct the error Grace made.

8. Write an equation (or several) for each problem. Use a letter for the unknown.

   a. One laptop costs \$480 and another costs \$35 less than the first.
      Mark bought both laptops. What was the total cost?

     Equation: _____

     Solution: _____ = _____

     The total cost was \$_____.

   b. Olivia earned \$16 for raking the yard and another \$16 for weeding. She had already
      saved \$40. How much money does she have now?

     Equation: _____

     Solution: _____ = _____

     She has \$_____ now.

Cumulative Revision, Grade 3, Chapters 1 - 4

1. Calculate.

a. $(18 - 5) - (3 + 6) =$ _____	**b.** $(300 - 50) - (80 - 30) =$ _____
c. $18 - 5 - 3 + 6 =$ _____	**d.** $300 - 50 - 80 - 30 =$ _____

2. Write each multiplication as an addition, and solve.

    a. 4×150 **b.** 3×50

3. Fill in the missing numbers.

a. ____ $\times 4 = 16$	**b.** _____ $\times 9 = 0$	**c.** _____ $\times 6 = 36$	**d.** ____ $\times 5 = 45$
____ $\times 8 = 64$	_____ $\times 3 = 27$	_____ $\times 4 = 36$	_____ $\times 2 = 18$

4. Calculate.

a. $8 \times 10 - 2 + 5 =$ _____	**b.** $6 + 7 \times (4 - 2) =$ _____
c. $3 \times 4 - 2 \times 3 =$ _____	**d.** $2 \times (4 + 4) \times 2 =$ _____

5. Solve. Write an equation or several for each problem to show your work.

    a. Tim has three rolls of string with 5 metres on each roll.
       What is the total length of the string on the three rolls?

       Equation: _____

       The total length is _____.

    b. A maths teacher bought four rulers that cost $6 each, and two kitchen
       scales that cost $22 each. What was the total cost?

       Equation: _____

       The total cost was $_____.

6. Continue the patterns:

a. $564 - 5 = $ _____

$564 - 10 = $ _____

$564 - 15 = $ _____

$564 - $ ____ $= $ _____

$564 - $ ____ $= $ _____

$564 - $ ____ $= $ _____

b. $888 + 12 = $ _____

$886 + 14 = $ _____

$884 + 16 = $ _____

_____ $+$ ____ $=$ _____

_____ $+$ ____ $=$ _____

_____ $+$ ____ $=$ _____

7. Draw an illustration that shows (proves) that $2 \times 7 = 7 \times 2$.

8. A word game has 24 letters arranged in an array, using six rows.
   How many columns does the array have?

9. One aeroplane ticket costs $267.

   a. Estimate the cost of three tickets using rounded
      numbers.
      They cost about $_____.

   b. Find the exact cost and check that your answer
      is reasonable.

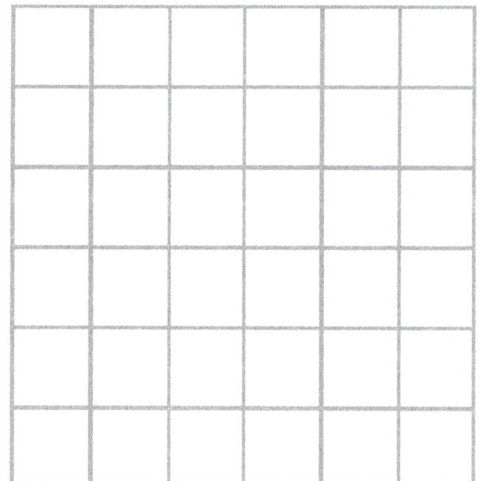

Puzzle Corner Place the digits 1, 2, 4, 6, 8, and 9 into
the boxes so that the equation is true.

___ ___ ___ $-$ ___ ___ ___ $=$ ___ ___ ___

Cumulative Revision, Grade 3, Chapters 1 - 5

1. Subtract in parts: break the second number into tens and ones.

a. 98 – 66

98 – 60 – 6 = _____

b. 54 – 26

54 – 20 – 6 = _____

c. 73 – 17 = _____

d. 62 – 25 = _____

e. 54 – 18 = _____

2. Subtract. Check your answers.

a.

$$\begin{array}{r} 9\ 0\ 4 \\ -\ 3\ 2\ 7 \\ \hline \end{array}$$ + _____

b.

$$\begin{array}{r} 8\ 1\ 2 \\ -\ 3\ 2\ 7 \\ \hline \end{array}$$ + _____

3. Study Jay's solution to the word problem. What is wrong with it?

Karen baked 30 cupcakes. She ate one. Her brother took two. Then her mother said she needed 2 cupcakes for each of the eleven ladies coming for afternoon tea. How many cupcakes are left after all that?

Jay's solution: 30 – 1 – 2 – 11 = 16. There are 16 cupcakes left.

4. How much time passes from 11:37 AM to 12:12 PM?

11:37   12:00   12:12

5. Write the time using the expressions "to" and "past".

a. 6:38

b. 3:56

c. 2:12

d. 7:43

6. Calculate in the correct order.

a. $7 + 4 \times 7$	b. $30 + 6 \times (6 - 6)$	c. $2 \times 44 - 8 \times 0$

7. The table shows how many bottles of milk a grocery store sold on different days.

Monday	Tuesday	Wednesday	Thursday	Friday	Saturday
37	32	28	24	42	34

    a. Round each number to the nearest ten and then estimate
       how many bottles of milk the store sold in total this week.

    b. Now calculate exactly how many
       bottles of milk were sold in total.

    c. Is your answer reasonable?
       How can you tell?

8. Solve. Write an equation for the problem using a letter for the unknown.

a. Mum had $200 when she went grocery shopping. She came back
   home with $78. How much did she spend in the store?

Equation: _____

She spent $_____.

b. The teacher gave each of the nine children 12 marbles to play a maths game.
   After the class, only 99 marbles were gathered back.
   How many marbles were lost?

Equation: _____

There were _____ marbles lost.

Cumulative Revision, Grade 3, Chapters 1 - 6

1. Ben and Joe travelled on a three-legged journey:

   (1) They took a bus from Lethbridge to Brooks.

   (2) A friend took them from there to Calgary.

   (3) Then they rode on a bus from there to Lethbridge.

	Airdrie	Brooks	Calgary	Edmonton
Airdrie				
Brooks	208			
Calgary	35	190		
Edmonton	269	425	299	
Lethbridge	241	155	209	503
Red Deer	117	305	147	155
Spruce Grove	288	453	322	33
St. Albert	289	445	319	16

a. Refer to the distance table above. Estimate the total number of kilometres they travelled.

My estimate: about _____ km.

b. Now calculate exactly the distance they travelled: _____ km.
You can use the grid above. Check that your answer is reasonable.

2. Round the numbers to the nearest ten.

a. $387 \approx$ _____	**b.** $992 \approx$ _____	**c.** $4 \approx$ _____	**d.** $345 \approx$ _____

3. Add in your head.

a. $49 + 13 =$ _____	**b.** $46 + 15 =$ _____	**c.** $25 + 39 =$ _____
d. $28 + 28 =$ _____	**e.** $26 + 27 =$ _____	**f.** $37 + 13 =$ _____

4. Write the time using the hours:minutes way.

a. 24 to 5	**b.** 10 past 11	**c.** 2 to 12	**d.** 17 to 1
_____ : _____	_____ : _____	_____ : _____	_____ : _____

5. Ben can walk to work in fifteen minutes. From Monday through Friday, how many minutes does he spend walking to and from work?

6. Multiply.

a. $5 \times 6 \times 2 =$ _____   **b.** $3 \times 3 \times 4 =$ _____   **c.** $5 \times 5 \times 2 =$ _____

7. Draw dots in groups to illustrate $2 \times 4 + 2 \times 7$.

   What single multiplication is equal to that?

8. Sylvia started exercising with an exercise video at 7:55 PM, and it ended 30 minutes later. At what time did it end?

9. How many hours is it?

from	8 AM	7 AM	9 AM	11 AM	10 AM
to	12 noon	1 PM	4 PM	11 PM	7 PM
hours					

10. Find the change by adding up.

a. You buy an item for $7.50. Find the change from $20.	**b.** You buy items for $38.45. Find the change from $50.
+[____] +[____] $7.50 $_____ $20.00	+[____] +[____] $_____ $_____ $_____
The change is $_____.	The change is $_____.

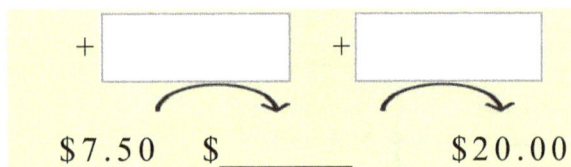

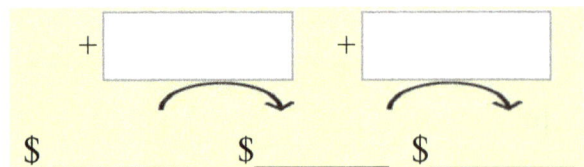

84

Cumulative Revision, Grade 3, Chapters 1 - 7

1. Add. Check your work by adding in a different order.

a. $372 + 49 + 32 + 286$	**b.** $8 + 76 + 478 + 304 + 128$

2. Figure out the pattern and continue it for six more numbers.

420, 431, 442, 453, 464,

3. Subtract.

a. $547 - 544 =$ _____	**b.** $109 - 99 =$ _____	**c.** $240 - 50 =$ _____
$800 - 792 =$ _____	$181 - 176 =$ _____	$560 - 260 =$ _____

4. Write multiplication equations to match the jumps on the number line.

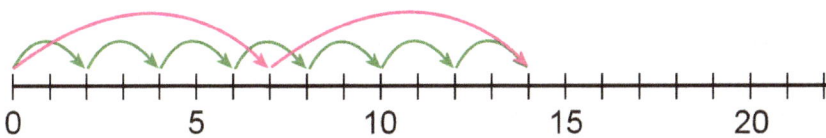

_____ × _____ = _____

_____ × _____ = _____

5. How much time passes from...

a. 2:26 a.m. to 10:26 a.m.	**b.** 8:15 a.m. to 8:40 a.m.	**c.** 4:25 a.m. to 5:00 a.m.

6. Write the time the clock shows. Below, write the time using "to."

a. _____ : _____

_____ to _____

b. _____ : _____

_____ to _____

c. _____ : _____

_____ to _____

7. Compare, writing < or > in the box.

a. 9018 ☐ 9180

b. 5000 + 600 ☐ 500 + 6000

c. 2387 ☐ 2378

d. 8000 + 50 + 2 ☐ 200 + 5000 + 800

8. Diana bought two dolls for $16.35 each and two teddy bears for $19.90 each.

a. Find the total cost.

b. She paid with $100. How much was her change?

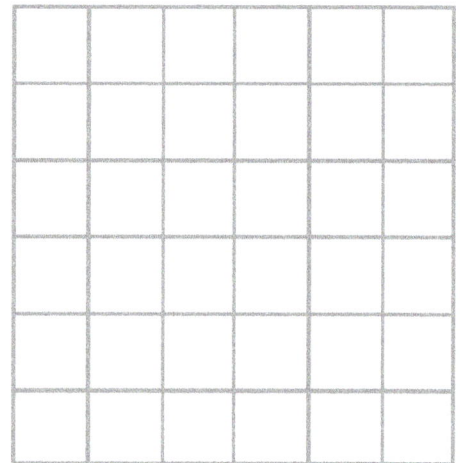

9. Find the missing money amounts. This is just like making change!

a.	b.	c.
$1.70 + _____ = $5	$2.30 + _____ = $5	$24.70 + _____ = $50
85¢ + _____ = $5	$3.70 + _____ = $10	$64.10 + _____ = $80

10. Find the missing numbers.

a.	b.	c.	d.
_____ × 2 = 24	48 = _____ × 12	6 × _____ = 36	60 = 5 × _____
_____ × 6 = 54	33 = _____ × 3	8 × _____ = 48	42 = 7 × _____

Cumulative Revision, Grade 3, Chapters 1 - 8

1. Write an addition and a subtraction equation to match the bar model.
   Fill in the missing parts.

a.

total ▭

| 780 | 130 |

_____ + _____ = _____

_____ − _____ = _____

b.

total ▭

_____ + _____ = _____

$2\,0\,5$ − $6\,5$ = _____

2. Multiply.

a. $5 \times 5 =$ _____

   $12 \times 12 =$ _____

   $7 \times 5 =$ _____

b. $2 \times 11 =$ _____

   $10 \times 56 =$ _____

   $4 \times 9 =$ _____

c. $7 \times 8 =$ _____

   $8 \times 12 =$ _____

   $6 \times 7 =$ _____

3. Trisha needs three whole weeks to write a report. If she starts writing on 3 November, when will she finish writing?

4. The science class starts at 10:55 and ends 50 minutes later. What time does it end?

5. Solve.

a. $120 + 5 \times 7 =$

b. $12 \times (9 - 2) =$

c. $(11 - 3) \times 3 + 5 =$

6. Write an equation using a letter for the unknown. Solve.

Ashley bought a gift for her mum for $39 and a bottle of water for $6, and now she has $25 left. How much money did Ashley have before buying these things?

Equation: _____

Solution: _____ = _____   Ashley had $_____ at first.

7. Children are watching videos. Write the starting or ending time. The length of the video is given below each exercise.

a.	b.
Start End 6:05 → ____ : _____ Video time: 40 minutes	Start End ____ : _____ → 7:40 Video time: 35 minutes
c.	**d.**
Start End ____ : _____ → 12:00 Video time: 15 minutes	Start End 2:45 → ____ : _____ Video time: 30 minutes

8. Carol bought six apples for $2.68, a carton of milk for $2.99, and a dozen eggs for $2.95. Calculate the total cost.

9. Fill in the missing numbers.

a. $40 \div 8 =$ ____ $72 \div 8 =$ ____	**b.** $30 \div 5 =$ ____ $56 \div 7 =$ ____	**c.** ____ $\div 6 = 6$ ____ $\div 5 = 5$	**d.** $72 \div$ ____ $= 6$ $45 \div$ ____ $= 5$

10. Find the number that the triangle stands for.

a. $1500 + \triangle = 2100$ $\triangle =$ _____	**b.** $5200 - \triangle = 4700$ $\triangle =$ _____	**c.** $\triangle - 2300 = 2300$ $\triangle =$ _____

11. Subtract. Check your answers.

a.   7 9 0 4
    − 3 2 9 7   +
	b. 5 0 1 2
	    − 3 2 7 +

Cumulative Revision, Grade 3, Chapters 1 - 9

1. Find the number that the triangle stands for.

a. $\triangle + 11 = 349$	b. $530 - \triangle = 320$	c. $\triangle - 161 = 500$
$\triangle =$ _____	$\triangle =$ _____	$\triangle =$ _____

2. Alex checked the price of a certain TV in four different stores.

a. Round each price to the nearest ten.

Now use the rounded prices and estimate:

b. About how much is the difference between the most and the least expensive TV?

	Price	Rounded price
Bob's TV Store	$525	
The Nerdy Store	$564	
Home Express	$632	
Lion Appliances	$599	

c. About how much more does the TV in Lion Appliances cost than the TV in The Nerdy Store?

3. Multiply.

a. $10 \times 14 =$ _____	b. $12 \times 12 =$ _____	c. $2 \times 4 \times 2 =$ _____
$11 \times 12 =$ _____	$11 \times 11 =$ _____	$5 \times 2 \times 12 =$ _____

4. a. Fill in the multiplication table of 9.

b. There were several special things about this table. What were those?

$1 \times 9 =$ _____	$7 \times 9 =$ _____
$2 \times 9 =$ _____	$8 \times 9 =$ _____
$3 \times 9 =$ _____	$9 \times 9 =$ _____
$4 \times 9 =$ _____	$10 \times 9 =$ _____
$5 \times 9 =$ _____	$11 \times 9 =$ _____
$6 \times 9 =$ _____	$12 \times 9 =$ _____

5. Solve. Write an equation for the problem, using a letter for the unknown.

a. Sheila and three other girls equally shared the cost of a taxi fare to the mall, which was $24. At the mall, Sheila bought a book for $18. How much did Sheila spend for the taxi fare plus the book?

Equation: _____

Solution: _____ = _____ She spent $_____ in total.

b. Julie has 16 golf balls and 8 tennis balls. She put the balls into bags, with four in each bag. How many bags does she need?

Equation: _____

Solution: _____ = _____ She needs _____ bags.

c. A meeting room has 6 rows of chairs, and a total of 54 chairs. How many chairs are in each row?

Equation: _____

Solution: _____ = _____ There are _____ chairs in each row.

6. Miriam spent three weeks in Florida, two weeks in Maine, and nineteen days in New Jersey. How many days did she spend altogether in the three states?

7. For each division, write a matching multiplication.

a. $32 \div 8 =$ _____

_____ × _____ = _____

b. $63 \div$ _____ $= 9$

_____ × _____ = _____

c. _____ $\div 7 = 5$

_____ × _____ = _____

8. Connect each problem to the correct answer. Cross out the problem that is not possible.

0×0 $5 \div 1$

1×1 **1** $0 \div 5$

0×1 $5 \div 5$

1×5 **0** $0 \div 1$

5×1 **5** $0 \div 0$

0×5 $1 \div 1$

9. Break each multiplication into two parts. Note that this can be done in several different ways. Lastly calculate the value of the expression.

a. 8×15	**b.** 6×21
___ × ___ + ___ × ___ _____	___ × ___ + ___ × ___ _____
c. 14×6	**d.** 4×23
___ × ___ + ___ × ___ _____	___ × ___ + ___ × ___ _____

10. Find the missing numbers.

a. $40 \div$ ___ $= 8$ $72 \div$ ___ $= 8$	**b.** ___ $\div 5 = 7$ ___ $\div 6 = 7$	**c.** ___ $= 48 \div 6$ ___ $= 72 \div 8$	**d.** $6 = 36 \div$ ___ $5 = 60 \div$ ___

11. The table shows how many fridge magnets Sarah sold at the flea market last Saturday.

a. Draw a bar graph from the data in the table. Decide the scale for the numerical axis. Include a **title** and **labels** for the bars.

Type	Number sold
Parrot	6
Toucan	15
Flowers	12
Palm trees	9
Ocean scene	18

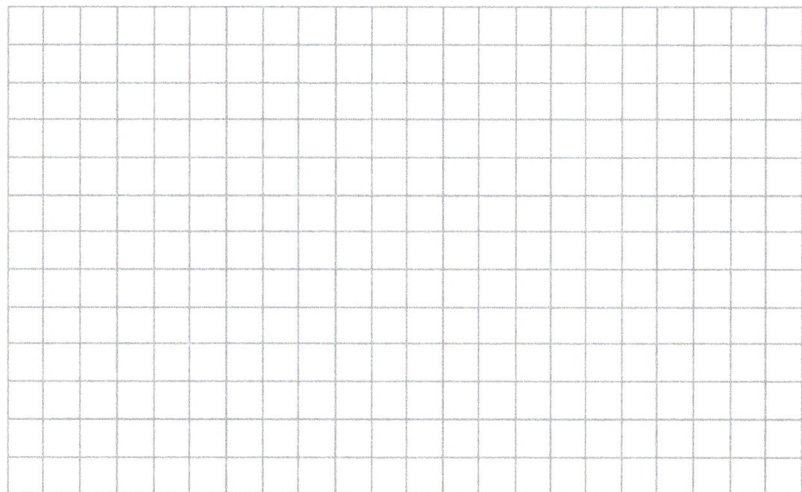

b. How many more ocean scene magnets did she sell than toucan magnets?

c. Which ones did she sell more of: the bird magnets (parrot or toucan), or all the other kind? How many more?

Cumulative Revision, Grade 3, Chapters 1 - 10

1. Add and subtract in your head.

a. $56 + 34 =$	**b.** $28 + 53 =$	**c.** $19 + 25 =$
d. $83 - 7 =$	**e.** $90 - 45 =$	**f.** $45 - 8 =$

2. Find the differences.

a. $40 - 38 =$	**b.** $93 - 88 =$	**c.** $640 - 631 =$

3. Write the multiplication & division fact families.

a.	**b.**	**c.**
_____ $\times$ 2 = 14	_____ $\times$ _____ = _____	_____ $\times$ _____ = _____
_____ $\times$ _____ = _____	_____ $\times$ _____ = _____	_____ $\times$ _____ = _____
_____ $\div$ 2 = _____	$35 \div$ _____ = 7	$42 \div 6 =$ _____
_____ $\div$ _____ = _____	_____ $\div$ _____ = _____	_____ $\div$ _____ = _____

4. Write the time using "hours:minutes".

a. a quarter past 1	**b.** a quarter past 5	**c.** a quarter to 8	**d.** a quarter to 12
_____ : _____	_____ : _____	_____ : _____	_____ : _____

5. How many minutes is it from the time on the clock face till the given time?

a. From	**b.** From	**c.** From
until 10:05	until 4:55	until 2:00
_____ minutes	_____ minutes	_____ minutes

6. **a.** List a measuring unit you use to measure the mass
of light items, such as an apple or a notebook.

b. List a measuring unit you can use to measure the mass
of heavy items, such as a refrigerator or a car.

c. List a measuring unit you can use to measure the
height of a TV.

d. List a measuring unit you can use to measure the
length of a room.

7. Write the weight using kilograms and grams.

a. _____ kg _____ g   **b.** _____ kg _____ g   **c.** _____ kg _____ g

8. Solve.

| **a.** $10 \times 61 =$ | **b.** $9 \times 80 =$ | **c.** $50 \times 6 =$ | **d.** $20 \times 12 =$ |

9. Solve. Write an equation(s) for each problem, to show the calculation(s) you do.

a. A car lot has 49 cars parked in an array, with seven cars in each row.
How many rows of cars are there?

Equation: _____

There are _____ rows of cars.

b. Shelly has three favourite cookie recipes. The first recipe makes 5 dozen
cookies, the second recipe makes 4 dozen, and the third recipe makes 2
dozen. How many cookies will Shelly have if she makes all three recipes?

Equation: _____

She will have _____ cookies.

10. Write the numbers in order from smallest to greatest: 6050 6553 5690 6055 6505.

_____ < _____ < _____ < _____ < _____

11. In the grid on the right, draw a rectangle that has an area of 12 square units. Then find its perimeter.

   Perimeter: _____

12. Fill in the missing parts. In the grid, draw a two-part rectangle that matches the equation.

   $3 \times (1 + 5)$ = ____ × ____ + ____ × ____

   area of the   area of the   area of the
   whole rectangle   first part   second part

13. Solve the problems.

 $7.25 $4.60    $16.90 $2.87

a. Matt bought butter, cheese, and one loaf of bread. He paid with $30. He got $2.25 in change. Was that correct?

b. Grace bought four cups of yogurt and paid with a $20-dollar note. What was her change?

Cumulative Revision, Grade 3, Chapters 1 - 11

1. The table on the right shows how many trees people from four different villages planted on Earth Day.

Silent Creek	120
Pine Valley	90
Riverside	150
Highland	60

Make a pictograph using a picture of a tree. Choose how many trees it signifies and write that under your pictograph.

Silent Creek	
Pine Valley	
Riverside	
Highland	

= _____ trees

2. Fill in the missing numbers.

a. $32 \div 4 =$ _____	b. $12 \times$ _____ $= 132$	c. $9 =$ _____ $\div 7$	d. $64 \div$ _____ $= 8$
$0 \div 5 =$ _____	$8 \times$ _____ $= 0$	$3 =$ _____ $\div 6$	$45 \div$ _____ $= 5$
$54 \div 6 =$ _____	$4 \times$ _____ $= 36$	$1 =$ _____ $\div 11$	$72 \div$ _____ $= 8$

3. Hal has two 5-cent coins, four 20-cent coins, and a 10-cent coin. He traded them to Bill for one coin of equal value. What was the coin?

4. Write a word problem that will be solved with the division $24 \div 6 =$ _____ .

5. Solve with mental maths.

a.	b.	c.
$1240 + 50 =$ _____	$1090 + 60 =$ _____	$3140 - 20 =$ _____
$8280 - 50 =$ _____	$9060 + 40 =$ _____	$7780 - 80 =$ _____

6. A maths book has a mass of 1 kg and an English textbook 1300 g.
   What is their combined mass? (Note: 1 kg is 1000 grams.)

7. A 20-litre bucket was full of rainwater. Marlene poured the water on five different plants in an equal manner. How much water did each plant get?

8. Sketch:

a. a rhombus	**b.** a quadrilateral with one right angle (the other angles are not right)

9. Ronny is building a clubhouse and the floor will be two metres by three metres.

   a. What will the area of the floor be?

   b. What will the perimeter of the floor be?

10. The picture shows Amanda's garden. She is going to plant potatoes in the smaller part, and different vegetables in the bigger part.

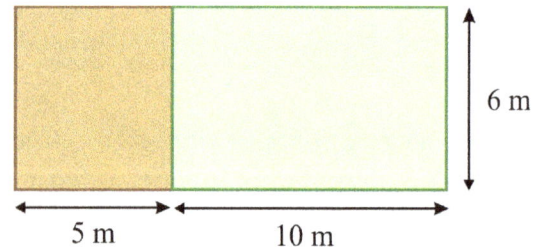

   a. Calculate the area that will be used for potatoes.

   b. Find the total area of her garden.

   c. Find the perimeter of the whole garden.

11. Draw a two-part rectangle with an area that is given by the expression $4 \times (1 + 6)$.

12. Show that $\dfrac{1}{2}$ is equal to $\dfrac{4}{8}$ using the number lines.

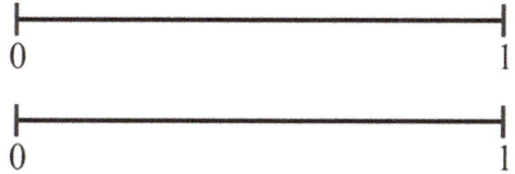

13. Some of the following equations are not true. Correct the ones that are false.

a. $\dfrac{7}{1} = 1$	b. $\dfrac{6}{1} = 6$	c. $\dfrac{1}{1} = \dfrac{4}{4}$	d. $\dfrac{2}{3} = \dfrac{4}{8}$	e. $5 = \dfrac{16}{4}$

14. The pictures show three fruit bars. They illustrate that $3 = \dfrac{3}{1}$.

   "Cut" each fruit bar into thirds. Now the illustration shows that 3 is equal to another fraction. Which?

15. Write the equivalent fractions. $\dfrac{8}{12} = \underline{\qquad} = \underline{\qquad}$

16. Compare the fractions.

   a. $\dfrac{1}{5} \square \dfrac{1}{6}$ b. $\dfrac{2}{8} \square \dfrac{9}{8}$ c. $\dfrac{3}{3} \square \dfrac{3}{4}$ d. $\dfrac{2}{1} \square \dfrac{7}{4}$

17. John says that 7/3 is less than 7/8 because 3 is smaller than 8.
    Draw a picture that will help John understand the truth of the matter.

18. Is this comparison valid? Why or why not?

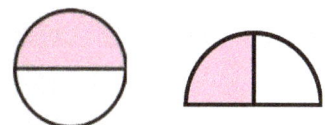

$$\dfrac{1}{2} = \dfrac{1}{2}$$

99

19. Subtract. Check by adding.

a. 3 2 1 3 – 1 4 1 8 +	**b.** 9 0 1 8 – 1 4 2 5 +	
c. 7 4 0 5 – 2 2 8 4 +	**d.** 3 4 4 0 – 9 5 1 +	

20. Find the perimeter of this shape in millimetres.

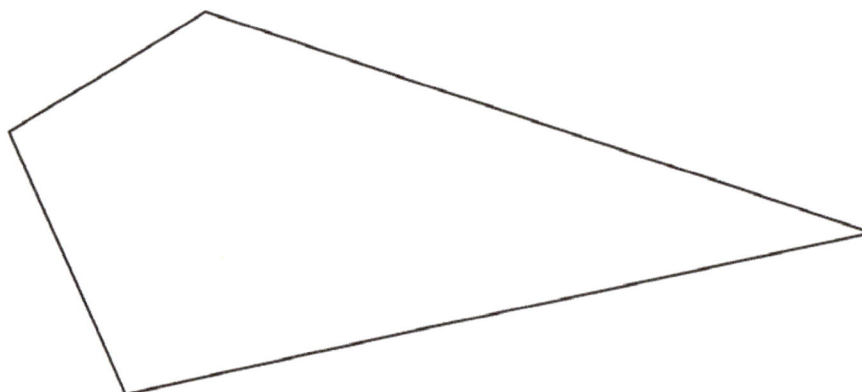